BLADESMITHING FROM SCRAP METAL

How to Make Knives with Leaf Springs, Cables, Railroad Spikes, and Files

TABLE OF CONTENTS

INTRODUCTION

I'm going to show you how you can forge blades using scrap metal, which you can probably find from a lot of different sources.

The one thing that you should pay attention to is the characteristics of scrap metal. Today, you can come into possession of high-quality carbon scrap steels and other alloys that the bladesmiths of long ago could have only dreamt of.

Notice the fact that I said 'carbon scrap steels'. You see, the main ingredient in steel is iron. However, the higher the carbon content, the better the metal will harden. If you find scrap made of stainless steel, then you may not be able to forge your blade readily. That is why, when looking for scrap, you should look for those that harden after processes such as heating or quenching.

But hardening is not the only thing you should be looking for when you are choosing your scrap. You should also pay attention to the type of steel that you are using. We will first start with some of the commonly used types of steel and which steel you should avoid.

Of course, we are going to delve into various scrap metals that you can use for forging as well, later on in the book.

On that note, welcome to the world of scrap bladesmithing!

FREE BONUSES FOR YOU

Before we start I have a something special for you.

To get the most out of this book, I have 3 resources for you that will REALLY kickstart your knife making process, and improve the quality of your knives.

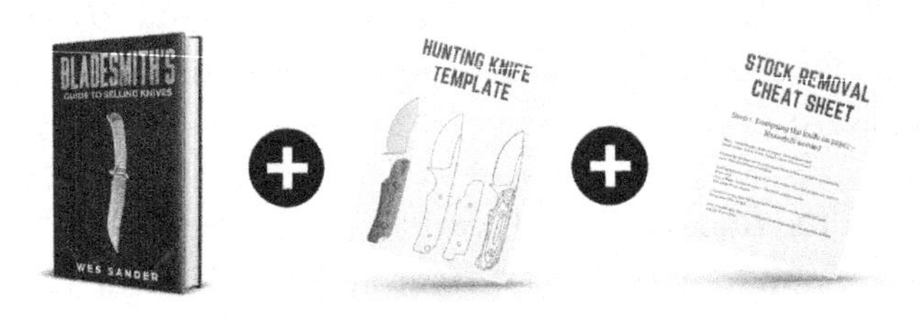

Since you are now a reader of my books, I want to extend a hand, and improve our author-reader relationship, by offering you all 3 of these bonuses for FREE.

All you have to do is go to **https://www.forge-hero.com/free-bonus** and enter the e-mail where you want to receive these resources.

These bonuses will help you:

1. Make more money when selling your knives to customers
2. Save time while knife making

Here's what you receive for FREE:

1. Bladesmith's Guide to Selling Knives
2. Hunting Knife Template
3. Stock Removal Cheat Sheet

Here is a brief description of what you will receive in your inbox:

1. Bladesmith's Guide to Selling Knives

Do you want to sell your knives to support your hobby, but don't know where to start?

Are you afraid to charge more for your knives?

Do you constantly get low-balled on the price of your knives?

'Bladesmith's Guide to Selling Knives' contains simple but fundamental secrets to selling your knives for profit.

Both audio and PDF versions are included.

Inside this book you will discover:

- How to **make more money** when selling knives and swords to customers (Higher prices)
- The **hidden-in-plain-sight** location that is perfect for selling knives (Gun shows)
- Your **biggest 'asset'** that you can leverage to charge higher prices for your knives, and **make an extra $50 or more** off of selling the same knife.
- 4 critical mistakes you could be making, that are **holding you back from selling your knife for what it's truly worth**
- The ideal number of knives you should bring to a knife show
- 5 online platforms where you can sell your knives
- 9 key details you need to mention when sell-ing your knives online, that will increase the customers you get

2. Hunting Knife Template for Stock Removal

Tired of drawing plans when making a knife?

Not good at CAD or any sort of design software?

Make planning and drawing layouts a 5-second affair, by downloading this classic bowie knife design that you can print and grind on your preferred size of stock steel.

Here's what you get:

- Classic bowie knife design **you can print and paste** on stock steel and start grinding
- Remove the hassle of planning and drawing the knife layout during knife making
- Detailed plans included, to ensure straight and clean grind lines

3. Stock Removal Cheat Sheet

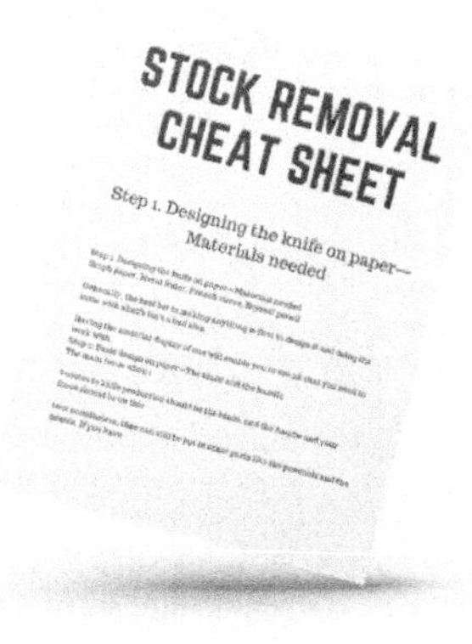

Do you need to quickly lookup the correct knife making steps, while working on a knife in your workshop?

Here's what you get:

- Make your knife through stock removal in just **14 steps**
- Full stock removal process, done with 1084 steel
- **Quick reference guide** you can print and place in your workshop

As mentioned above, to get access to this content, *https://www.forge-hero.com/free-bonus* and enter the e-mail where you want to receive these 3 resources.

DISCLAIMER: By signing up for the free content you are also agreeing to be added to my bladesmithing e-mail list, to which I send helpful bladesmithing tips and promotional offers.

I would suggest you download these resources before you proceed further, as they are a great supplement for this book, and have the potential to bring an improvement in your results.

CHAPTER 1: TESTS FOR CHOOSING THE RIGHT SCRAP STEEL

SNAP TEST

When it comes to steel, you can check the carbon content of the material by using the snap test. You can also check the grain size.

In this test, take a small sample of the steel that you are planning to use and forge it to about 2-3mm thick. Ensure that it is about 3cm long and about 1cm wide. Once you have gotten the measurements right, heat the entire piece of metal to above non-magnetic levels. Then quench it in water to harden it to its full potential.

Next, take the metal and clamp it tightly using a tool like a vise. Take a hammer or a large tool that is capable of breaking the piece of metal (I recommend a hammer and in any case if you are working with metals it is important that you have a hammer around anyway!)

Strike the piece of metal with your hammer until it either bends to a 90-degree angle or breaks off completely.

Here's what the test results are if either of the above situations occur (bending or breaking):

- If the metal breaks without bending, then the metal has a high percentage of carbon.

- If it bends without breaking, then it has a relatively low percentage of carbon.
- If is bends partially and then breaks, then it is probably somewhere in the middle (an average percentage of carbon).

As mentioned above, it is not just the carbon level that you can check with this test. You can examine the steel and discover the grain size, too. The grain size gives you more information about the properties of the steel, such as the yield strength of the metal.

Usually, the smaller the grain size, the stronger the steel. This is because when the grain size is large, there is a greater chance for certain 'dislocations' or breakages to occur.

HARDNESS TEST

A snap test sounds convenient. But is there a test to gauge the hardness of the steel? Turns out, there is. Let's look at the Rockwell steel hardness test.

One of the characteristic features of the Rockwell test is that it is relatively easier to perform and has a higher degree of accuracy than other tests. You can also use this test on almost all forms of metal, which makes it useful for materials other than steel as well. The first thing that you need to do is find yourself a hard steel ball or a diamond tip that acts as an indenter.

You then need to perform a preliminary test. A certain amount of force is applied with the indenter to the piece of metal wherein it begins to break through the surface. The force leaves an indentation in the metal, which is then measured. This indentation is known as the baseline depth.

Next, an additional load is then added. This produces a deeper indentation, which is then measured as well. This particular indentation is known as the final depth. Once you have discovered both the baseline and final depth, you find the difference between the two values.

The resulting measurement is the hardness level of your metal.

But how can you find out if your steel is ready for you to forge it into a knife? Well, you use another simple, but effective, technique. All you have to do is heat the metal to its non-magnetic temperature. Once the metal reaches the temperature, you then quench it. If you discover that it hardens, then your metal is suitable for bladesmithing.

THE RISKS WITH SCRAP METAL

Finding scrap metal is easy and definitely works well for beginners and those who do not want to spend too much time obtaining the right materials.

But scrap metals come with its own challenges. Here are some:

1. These days, no two coil springs are the same. With the number of different vehicles on the roads, each vehicle manufacturer uses a unique set of coil springs. This means that different coil springs should be heat treated differently.
2. Another challenge you might find is that once you finish working on scrap metals, you might end up noticing hairline cracks. The presence of these cracks ruins the presentation of the entire tool.
3. Finally, the cracks can also reduce the durability of the metal and might even cause the metal to break entirely.
4. You are always going to have to expect the unexpected. When using scrap metal, you can never be certain of what you are working with.
5. Make sure that you know which steel you are using to make the knife. This will allow you to heat treat the knife properly.

CHAPTER 2: LEAF SPRING

Most leaf springs are made from 5160 steel. Refer to the properties section under Chapter 9 to find out more about 5160 steel.

NORMALIZING

The next process that you are going to focus on is normalizing. It is good to normalize after you forge your knife because you can refine the grain of the steel. And refining the grain is important because it allows the hardness of the metal to increase.

1. For the normalizing process, heat up the knife to the yellow temperature range again; making sure that it is done slowly and uniformly.
2. Hold the knife at the temperature mentioned above for a period of time. You should ideally be normalizing your knife for one hour per one inch of thickness. This means that if your knife is two inches thick, then your normalizing period would be two hours.
3. After the normalizing duration, take out your knife and let it cool in the open air. Allow it to cool until it reaches room temperature.
4. Remember that normalizing is all about uniformity and you should ensure that the entire

blade region of your knife is given a uniform normalizing treatment.

FORGING

Let's go with leaf springs to start things off.

1. You are initially going to get a lot of metal to work with. For this reason, you are going to use a hacksaw to cut off a small portion from both ends so that it makes it easier to work with the metal. Ideally, you can remove about one inch from each end.
2. The leaf springs you use will probably have a slight arc to them. You first have to take your hammer or any other heavy instrument and then hit the spring until it turns flat.
3. The best way to do this is not by striking the metal the just as it is. If you do that, you will probably end up hitting it a lot of times before you see any change at all.
4. Instead, first place the metal into the forge and heat it. Let it enter the yellow temperature range. The ideal forging temperature for leaf springs is in the yellow range.
5. Once that is done, place your piece of metal into an anvil and flatten it. Use your hammer for this. Strike the metal as much as possible until it is flattened. If you need to, you can place the metal

back in the forge to heat it up again so that you hammer more easily.

6. Once the metal has been flattened, draw out the template of your knife on the spring.
7. Once that is done, use your hacksaw to cut out the shape of the knife.
8. Next, we are going to add the bevel. Take the knife over to the grinder and begin to slowly add the bevels. Take your time with this process so you can refine the shape of the knife.
9. After you have forged the knife, make sure that you have removed the scale. Once done, mark the holes that you would like to create into the handle of the knife and then use a drill to make the actual holes.
10. Grind out any bevels in the knife and give it a smooth finish.

HARDENING

1. Once again, you are going to heat the knife to its transformation range (the yellow range). At this point, you can check to see if the knife is magnetic or not. Simply take your magnet and bring it close to the metal.
2. If the metal is still magnetic, then heat it a bit more and check again. When the knife is truly non-magnetic, then take it out of the forge and get ready for the quenching process.

QUENCHING

Most people think that they have to invest a lot into the quenching process. But the reality is that you actually don't have to.

You can get yourself food-grade quenching oils. They are a lot cheaper than commercial quenching oils. They are also easily available. You just need to drop by your local store to find them.

Two of the most common oils used for quenching are canola and peanut oil. The main reason for choosing these two oils is the high flash point that they have, which is ideal for the quenching process. Before you quench the metal into the oil, make sure that you preheat them anywhere between 120-130 degrees Fahrenheit. Quench the knife for about 11 seconds.

TEMPERING

You temper the steel immediately after hardening. When the knife cools down, you reheat it to a temperature between 302-752°F.

For the leaf spring, you can use a regular oven for the tempering process. All you have to do is preheat the oven to 375°F. Let the oven preheat completely before you place the steel inside. Once the metal is inside, keep it there for about three hours at that temperature. You can place your metal directly on the oven rack or you can put it on top of a baking sheet.

Don't have an oven? Then you can use a toaster if your metal can fit inside. Or alternatively, you can use the toaster if you are planning to prepare some cookies in the oven!

If you have neither of the equipment above or if they are both engaged, then the best way to temper the steel is by using a blowtorch. You need to focus the flame from the blowtorch on the area that you want hardened. You will know when the tempering process is complete when the blade turns a blue color.

Finally, you are ready to grind the bevels and fix the handle. Once you have attached it, sharpen the knife until you give it the edge that you want. To finish off the process, polish the knife and the handle to add that extra special gleam to it.

Your knife is now ready.

CHAPTER 3: CABLE DAMASCUS KNIFE

Working with a cable is not difficult. You might think it is a complicated process, but what you really need to understand is that you have to be patient with it.

The first thing to remember is that you should not pick stainless wire for your knife. This is because when you are working with the welding process, you might find out that stainless wire does not get welded into the billet. Additionally, it is also toxic to work with stainless wire.

Here is the process you need to follow for a cable (Damascus knife):

1. Start off by either arc-welding the cable or tying up the end of the cable. This will prevent the cable from unlaying, wherein the strands of the cable begin to come off.
2. Place the cable into tar or oil; making sure that it is soaked completely.
3. Take it out of the tar or oil and then place it into the forge. There is often a debate about when the metal is ready after placing it into the forge. You should let the tar or oil burn out completely.
4. Once you have noticed the tar and oil burn out, use your tongs and flux the metal. Ideally, you should be using borax for this process. Make sure you flux the metal liberally, with particular focus

on the center. Try and get the flux to soak into the metal. The main purpose of the flux is to prevent oxidization. You also ensure the metal is fluid and remove any other form of impurities.

5. Next, heat the metal again. Take it out of the fire, brush the metal, and then flux it again. Take it back into the forge and heat it again.
6. At this point, you are ready to hammer the cable. Take out your hammer and pound on the cable lightly. Once done, place it back into the forge and bring the color of the cable to roughly the same color as inside the forge. When you feel like the color has been attained, take it out of the forge and get ready to weld it.
7. To weld, make sure that the ends of the cable are brought together. Weld it until it has a nice rectangular shape. There are two ways of welding the ends together:
   - You can take the cable to its welding temperature. Then transfer it to a vise and twist the ends together.
   - Or, you can simply use the hammer for the ends.
8. It is time to forge the cable. When you are forging, remember that when you first bring the hammer to the cable, your blows should be soft and precise. Do not strike it too hard or you might risk damages on the cable. While you are hammering, roll the cable around and make sure that your

blows strike the cable evenly from one end to the other.

9. Once you are done hammering, run it through the flux process again. Use a wire brush on the cable's surface, making sure you brush it all over. You only need to run the brush a couple of times on the cable.
10. Next, grind out the bevels.
11. Now you are going to heat-treat the blade. We start with the normalizing process. We are going to use a kaowool on the blade, which is essentially a type of insulator. Cover the tip of the blade with a kaowool and place the knife into the forge. This way, you ensure that the entire blade receives the heat treatment evenly. Heat up the metal to around 1,300°F and then place the metal inside the forge for an hour. Come back and allow the metal to cool in air for an hour. Repeat this once more.
12. Once you have finished the normalizing process, make sure that you remove the kaowool so that you can get it ready for the quenching process.
13. Once you have removed the kaowool, dip the blade into the quenching liquid of your choice.
14. Your heat treatment is done and you are ready to attach the handle.
15. Fix the handle.
16. Sharpen the knife.
17. Polish the blade and handle.

CHAPTER 4: RAILROAD SPIKE

When you are getting started with forging or if you are still getting used to the process, one of the ways that you can forge is by using a railroad spike. This is ideal for those who are just entering the world of bladesmithing or those who are simply out of touch with the various techniques.

One of the common misconceptions about railroad spikes is that they are a material with high carbon content. It is true that bladesmiths around the world have used railroad spikes to craft a blade, but they do not have the carbon content that most think they do. The reason that bladesmiths enjoy using them is that they are almost readily available and usually for free.

The misunderstanding about their high carbon content comes from the fact that there are some railroads marked with the letters 'HC' for 'High Carbon.' But what that means is that though there is a high carbon content for the manufacturing of railroads, it is not high enough for blade making.

What is high carbon for railroads is low carbon for knifemakers.

Here is something to remember:

Based on specifications set forth by American Railway Engineering Association, railroad spikes can be divided into two categories: low carbon track spikes that are used on those sections of railroads that are straight, and high carbon steel track spikes used on switches and curves. The association's guidelines mention that, low carbon spikes may contain no more than 0.12% carbon and "High Carbon" spikes may contain no more than 0.30% carbon.

But how does this affect knife makers? Well, to make a good knife, the blade needs to contain anywhere between 0.85% to 1.5% carbon.

That is three times or more than the carbon content contained in railroad spikes.

But why use such low amounts of carbon in railroad spikes? Aren't they supposed to be harder in order to keep the rail tracks together? The reasoning behind the low carbon is that the spikes can bend a little. A bent spike can hold the rail track but a broken one (due to high carbon) will not be able to hold the track together for a long time.

Thus, this is why you cannot use a railroad spike for making tools that you will use regularly, such as knives. Railroad spikes do not have a high enough carbon content and they easily suffer cracks on the surface.

However, it is easy to start with railroad spikes and get yourself acclimated to the various processes involved in forging.

That said, if you have received an order from a customer to make a knife, then you should not be using railroad spikes for it.

1. When working with railroad spikes, it is important to note that many of them do not survive the hammering process if they are heated to temperatures below 1,200°F. What do I mean by the fact that they don't survive? Well, they are easily prone to breakage and cracking. For this reason, you should start the process by heating the spike to temperatures between 1,200°F to 1,500°F.
2. Here's something you *can* do with the railroad spike. Once you have heated the spike to the temperature mentioned above, you can add designs to it if you wish. You can do this by placing the spike between a vise and twisting it. When you are twisting, make sure that you only focus on doing half rotations or full ones so that you can keep the point of the spike pointing in one direction.
3. Heat up the spike again. This time forge it into the shape of the knife that you would like to attain. The longer you want the tip to be, the more times you might have to put the knife in the forge to attain the desired effect.
4. It's time for the grind. When you reach this process, you are going to ensure that you clean off the scale and grind away whatever you do not want on the knife. Take your time with this step so that you have results that you are satisfied

with. You should also make sure that you grind a bevel and an edge.

5. Your knife is looking good and you are almost ready to finish the process. Some people decide to polish the blade at this point, but it is highly recommended that you do not do so. If you are punching holes, make sure that you do it now before the heat treatment process. We are going to heat treat the blade at this point so let's get into that process.
6. First, gradually heat the spike in the forge until it reaches the orange range. You can check and see if it has reached the non-magnetic stage at this point. If it hasn't, then heat the spike a little more until it reaches the non-magnetic point.
7. Once that is done, take the spike out of the forge and quench it in the liquid of your choice.
8. Now we are going to use a method to see how smooth your knife has turned out. Use a file and run it against the knife. If the file slides along your knife, then you have worked through the hardening process wonderfully.
9. Once you have quenched the spike, lift it out of the liquid and temper it. As we had noticed in Chapter 2, you can temper it in the oven by preheating the oven to anywhere between 350-400°F. You can also use your toaster if the knife can fit into it. Or you can make use of the blowtorch as mentioned above. The choice is entirely yours.

10. Now, you can add the handle on the knife. When you are going to move the knife to the scales, make sure you wipe off the inside of the scales before you apply epoxy. This ensures the scales will be dry.
11. Right before the final step, make sure that you use acetone to clean up the knife. You can then apply a coat of varnish.
12. For the final step, you just have to polish and run the knife through one more stage of sharpening.
13. With that, you have completed working on the knife.

CHAPTER 5: COIL SPRING

A coil spring is an important component of any vehicle. It is an elastic component (though not *that* elastic), which helps the vehicle absorb shocks. It usually takes in the shock and gently puts it back in the direction of the ground in such a manner that it prevents any damage to the vehicle itself.

But when they are not used in vehicles, they can be used to make some rather decent knives. It is true, however, that you might have to make thinner knives when you are using them.

Forging a knife from coil spring can be a bit of a challenge. But despite the challenge, coil springs are still an ideal steel for making knives. One of the more difficult parts of working with a coil spring is trying to straighten it. People often try numerous ways and give up in exasperation.

But luckily, there is a good technique to easily straighten out the coil without tearing out your hair in frustration. It has to do with the anvil. But first, let us check and see if the coil is ideal for forging.

This is an important step, because sometimes you may eventually find small hairline cracks that you would not have seen earlier. So go ahead and examine your metal carefully to check for those. You should ideally aim to get a coil spring that does not have many of the aforementioned cracks. Coil springs of older vehicles are preferred because they have 5160 steel in them, which is ideal for knifemaking. To understand how old the vehicle has to be, check and see if it was manufactured 10 years before. If so, then you have yourself some nice steel to work with.

One of the important things to remember is that you need to plan out the knife that you are going to make when you get your hands on the coil spring. You may not have enough metal to make a broad blade. Instead think about something narrow, so that you can forge accordingly.

Your first step is to use a small part of the coil and heat it. Ideally, choose an end of the coil for this purpose and if your coil is ready for forging, you can use the other end to work on your knife.

1. Heat the coil until it reaches the orange range. Then quench the coil in the liquid of your choice. Check and see if the coil hardens. If it does, then you can use it for knifemaking. If it doesn't, you will unfortunately have to find another coil.
2. To straighten out the coil spring, make sure that you cut it off short. Once that is done, you can use the hardy hole that comes with the anvil to hold the spring in place while you straighten it. First, take the entire coil and cut off a half portion of it using an electric saw in such a way that it resembles the shape of a horseshoe. Next, place the cut out coil into the forge and heat it until it

enters the orange range. Place the hot coil into the hardy hole of the anvil and bend it using your tongs until it straightens. And that's all there is to it!

3. Once it is straightened, you can then use your hammer to flatten it more. At this point, you should be looking to flatten it in such a way that you start noticing the shape of your knife. You might need multiple turns at the forge to get the right shape, but eventually, you will be able to make it work.
4. After flattening out the coil and straightening it into the shape that you want, you can then cut off any additional length from the metal. A simple saw will do the trick (since at this point it is not too thick to provide too much resistance).
5. In your hands, you now have the crude shape of your knife. We are going to work on it until we get it to the finished product that we desire.
6. Now it is time for the forging process. Place the knife into the forge and wait until heats up to the orange range. Take out the knife and then use your hammer to make the knife shape.
7. Take your knife to the grind next and add the bevel to it. Smoothen out the surface and remove any markings or remnants of the coil springs from it.
8. Take out your trusty drill. We are going to drill holes into the handle. Punch in the desired number of holes into the handle based on the design

you are going for. When you have finished adding the holes, your knife will actually look like it's finished. But the real work is far from over. We are now going to send it to the forge.

9. If you would like to see whether the steel is okay and if there is anything wrong or amiss about it, try and etch it using ferric chloride. One of the important uses of the ferric chloride is to remove any phosphorus and to reduce hydrogen sulfide. Essentially, you are making sure that the quality of the steel that you are working with is up to your preference.
10. Once you have ascertained the quality of the steel, place it into the forge and allow it to heat until it reaches the orange range. At this point, we are only focusing on heat treatment. So as soon as your metal has reached the required temperature, shift it immediately into a quenching liquid.
11. You need to be sure get the right temperature for this process. Place the knife into the forge until it reaches around 1,500°F. Once the temperature has been reached, take out the knife and dip it into a quenching liquid. For this metal, ideally I would recommend that you use canola oil. Make sure that you preheat the canola up to a temperature of 400°F.
12. You need to heat treat the metal for two 2-hour cycles. Essentially, you are going to follow the above steps and ensure that the knife receives a heat treatment for the full two hours. Once done,

you have to repeat the process again, making sure it once again receives the heat treatment for two hours.

13. With that, you have completed the heat treatment for the blade.
14. Once again, take the knife to a grinder and make sure that you remove the scaling from the heat treatment. Use the grinder or sharpening stone to bring the edge of the knife to your desired state. Smoothen out the surface if necessary.
15. With that, you are done with the coil spring knife. The final step basically involves merely adding the handle to the knife.

CHAPTER 6: KNIFE FROM A NICHOLSON FILE

When you are using a file, you should keep in mind a few important things when getting started. The first is that you should ideally aim to work with a file that has at least 0.9% carbon. Along with the carbon content however, you need look for additional mixes. You should have a little manganese and trace amounts of chromium, vanadium, and tungsten. With these combinations, you are going to get a metal that has a good level of hardness. Luckily, with a Nicholson file, you are getting a truly hard steel.

Why is it such hard steel? That is because a file is mainly created in order to cut through metal.

Nicholson files have a sense of consistency. This means that no matter what Nicholson file you use, they are always made of the same metal: **1095.**

In order to make a knife using a file, you must make sure that you soften the file before you proceed with the knife-making (since it is hardened to be used as a file). However, you can still make a knife without the softening process. Which brings us to the following two options:

- The first option is that you can directly use the hard file to make your knife. The result will be rather crude and it might look aesthetically unpleasing. But it is the fastest and simplest method to make your knife. You could try this method if you would like to practice a bit, but there are also other ways you can practice making a knife, as we had seen earlier chapters.

- The second option is to subject the file to the softening process, which in this case is annealing. With this process, you can soften the metal and create a knife that not only looks visually pleasing, but also looks more like a knife than a prison shank (which might be the case if you are using the first method). Obviously, you get a better knife using this process, so let's go ahead and expand on option number two.

We have to start off with the annealing process.

1. Use your forge and heat the blade to a temperature of 1,350°F. We are going to use the magnet trick to check and see how hot the metal really is. If the magnet sticks to the surface, continue heating it until you have reached the non-magnetic temperature. If you like, you can also use a torch to heat the surface of the file evenly.
2. Eventually, you will notice that the magnet stops being attracted to the metal. At this point, you might notice that the color of the metal is red hot. Keep the metal at the red-hot range for about 2-3 minutes (after making sure the metal is non-magnetic).

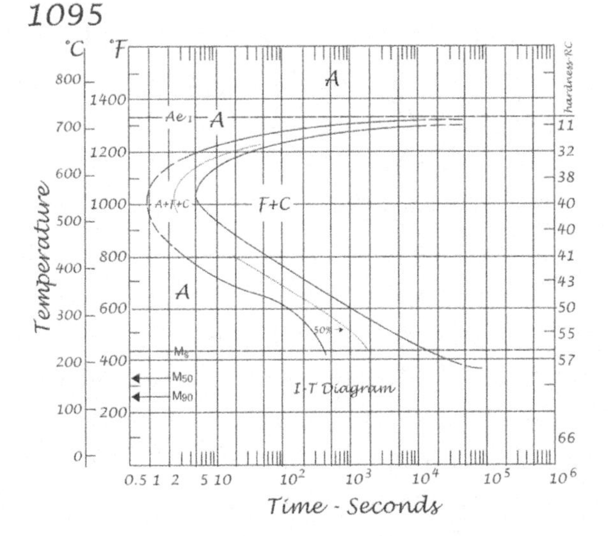

3. Allow the blade to cool off in the air. You should be able to watch as the glow begins to reduce. When it reaches a dull red glow, quench the blade. You can use Vermiculite. Wood ash is another option for you to choose from. Make sure that whatever option you choose, it covers at least six inches all the way around. The more it covers, the better your knife will turn out to be.
4. Once you have quenched it, the blade will be left in a softened state.
5. Your steel has now been properly annealed is ready to be treated all processes required to turn it into a knife.

6. Take out your trusted drill and make the holes into the tang. Use as many holes as required, based on your preferences.
7. Head over to the grinder and shape the profile of the knife. Whenever you notice the file getting hot, stop your work and dip the knife into water to cool it. You should keep the temperature of the file under 450°F. This might be difficult to manage but the trick is to cool the knife immediately whenever you notice that it is getting too hot. Do not wait to see how much heat you can handle. If you are not sure when to dip it into the water, just go ahead and cool the metal even if it gets just a little hot.
8. Another way to notice if the knife is getting too hot is to look for blue spots. When you notice these spots, it means you are heating up the knife too much. Try to avoid reaching this point, but use them as emergency limiters to ensure that you do not ruin the knife any further. If you end up having too many blue spots, the knife might become too soft in the final grinding process. Hence, the importance of giving extra attention to this during the process to avoid any complications in the future.
9. When you have created a profile of the knife, begin grinding the bevels. Eventually, your knife will look more like the shape you want. When you reach this point, smoothen out the knife and refine the shape as much as you can. (The guide

to doing a flat grind on your blade is at the end of this book)

10. Harden the steel to just above its non-magnetic temperature.
11. For the next step, you are going to temper the steel. To do this, we can use the kitchen oven. Simply preheat the oven to a temperature of 450°F. Place the metal into the oven and allow it to heat up. Let the metal remain in the oven for about two hours.
12. Take out the metal and allow it to cool down to room temperature. When it is cool, take it back into the oven and let it heat for another two hours. This time, when the two-hour cycle is complete, you are going to cool it down with water.
13. We are going to use sandpaper to remove any marks from the knife. Take out some 100-grit sandpaper for this process and begin sanding the knife. Sand both sides of the knife and work your way to all the surfaces, including the handle. You are going to get a well-shaped and smooth knife from this. You will notice that the edge is not sharp yet, but that is alright. We will come back to sharpening it later.
14. Now, switch to a 220-grit sandpaper and repeat the process. Make sure that you start with both sides of the knife. Finally, use the 400-grit sandpaper. Your knife will start giving out a nice shine.

15. With that, your blade is ready to receive a handle. If you have already prepared your handle, you can attach it to the knife at this point.
16. At this point, you can begin to sharpen the blade. Take out a sharpening stone and work on the edge of the knife until you can see it reach your preferred sharpness.

CHAPTER 7: SCRAP DAMASCUS

In this method, you will be trying to take different pieces of metals and then piling them one on top of the other.

For this purpose, you are better off working with band-saw blades and files.

Let's look at each of the materials that you are going to use. When using bandsaw blades, remember that you might come across different saws that have different qualities. You should first be looking at the band saw blade diameter. This tells you which bandsaw you can use for which blade and more importantly, which band-saw can be used to cut the blade.

If you choose to make blades that are thicker, then they can withstand more strain from straight cuts. On the other hand, when you are using thinner blades, they are perfect for some of the lighter work that you do. But the best way to choose the right bandsaw is by trying to figure out the width of the knife that you are creating.

With this in mind, you can use the handy table below to get an idea of which bandsaw is perfect for your blade width.

Wheel Diameter (inches)	Blade Thickness (inches)
4 – 6	.014
6 – 8	.018
8 – 10	.020
11 – 18	.025
18 – 24	.032
24 – 30	.035
30+	.042, .050, .063

Using the chart above, you can find out the ideal band-saw for the right blade.

1. The next part is cutting the bandsaw and the file so that you can allow them to be stacked on top of each other. We are going to be using a band-saw to cut the bandsaw that you have in your hands. Refer to the table above to choose the right bandsaw for cutting the file and the bandsaw.
2. Make sure that you cut all the pieces to have the same dimensions.

3. Next, take out the grinding wheel and then smoothen out the surfaces of the files and the bandsaws. Remove as much of the marks as possible so that there are no chances of cracks forming on the surface during the forging process.
4. Once you have done that, you need to get the plates to 'stick' together. It is only then that you will be able to drop the metal pieces into the forge and then continue with your project. To make them stick, you need to weld the pieces of metal together.
5. Start by welding the corners of the metal. This will allow you to put them together without them easily falling apart, especially when you add them to the forge.
6. Once the welding process has been completed, you are then ready to use the forge. Place the piece of metal you are holding into the forge. Wait for it to heat up. As the metal continues to heat, take it out using the tongs and flux it with borax. Alternate between heating and adding flux until you notice the metal reach the yellow range.
7. When the metal is glowing yellow, you can take it out of the forge and then gently use the hammer to set the weld in. What this means is that you are hammer the metal so that all the pieces begin to like one large block of metal rather than smaller pieces of metal.
8. Once you have completed this process, you will notice that the original weld that you had placed

on the metal is still there. It will look like small mounds on the corners of your metal. The best way to get rid of these mounds is by using a saw. Make sure that you are especially careful while trying to cut the welds or you might accidentally cut of a piece of metal or maybe even add a little dent in it.

9. When you have finally removed the welds, you need to place it back into the forge. Allow it to enter the yellow range of heat and take out the metal. Hammer it until you do not notice any visible signs of the welding process or any indication that there are different metals stuck together.
10. Next, head over to the grinder and take out any unevenness or marks that are on the metal. Continue with the grind until you have a nice even shape in front of you. At this point, your piece of metal should look like a rectangular block of solid steel without any indications of indentations, lines, or cracks.
11. If you want to create your knife from this metal, you only have to follow the steps below. You are basically going to go through the typical forging process.
12. Head over to your forge and place the steel into it. Wait until it turns hot enough that it enters the yellow range. Once it is in the range, you can take it out and begin hammering it. Get the shape of your knife into the metal. You might have a rudimentary design of your knife at this point, but

that it just fine. We are going to be adding in more details once we are done with this step.

13. Head over to the grinder and begin getting your shape fine-tuned. Once you have done that, you can then subject it to the heat treatment process.

CHAPTER 8: ANVIL FROM A RAILROAD TRACK

One of the best parts about working with an anvil is the fact that you actually only need a few simple tools to complete the process. This might be contrary to what most people believe. In fact, one of the common misconceptions about making an anvil is that people think they are going to have to invest in a lot of equipment to work on the railroad track in order to finally change it into the shape that they desire. In reality, you are not going to be involved in such a complicated process, as we will notice below.

When you are working with railroad tracks you have to make sure that you are choosing the right one.

There are a few different types. First, you have a railroad track that is made for large trains and locomotives. And then you have the tracks that are made for smaller carriages. What kind of smaller carriages? Well think of those wagons that are used in mines.

Since you are making an anvil, you need to have a railroad track that is on the bigger end of things. Which is why you should be looking for used railroad tracks from trains.

When you are looking for tracks that you would like to work with, you should go for the 100-pounds/yard tracks. As for the length of the rail, you should go for one that is a foot long.

When you have the track with you, you should focus on creating a hardy hole (which we saw earlier as being used to straighten a piece of metal). You might start off with a small hole and once you have made it, you are going to expand it even more. In order to achieve that expansion, you are going to make use of a large punch or chisel for the job. Below is a diagram of what we will try to achieve with the railroad track.

1. Take out your chisel and using the hammer, punch through the hole until you see it getting bigger.
2. Now you have a foot-long railroad track with you. But it doesn't actually look anything like an

anvil now, does it? For that reason, we are going to trace the shape of the anvil on top of the railroad track.

3. Take out a chalk and face the anvil lengthwise. You are now facing the width of the anvil. Find the midpoint along the width. We will call this point 'Point A.' Essentially, you need to focus on one end of the railroad track, find the midpoint, and mark it. Once done, draw a line that connects the midpoints of one end of the railroad track to the other end of the railroad track.
4. Once you have drawn the line, you now need to find the midpoint on that line. Once done, draw the second line across the midpoint of the first line, such that they connect the two sides of the anvil. You can name the point on one side as 'Side A' and the one of the other side as 'Side B.' Basically, you now have a cross shape on the top face of the railroad track and you have further split the top face into four equal sections.
5. Now comes the tricky part. Start from Point A and trace a curved line to Side A. You can choose to draw a straight line; but in order to get the right shape of an anvil, I prefer a curved line. Next, draw another curved line from Point A and connect it to Side B.
6. You now probably have a rough triangle shape on one end of the top of your anvil. The next step is fairly easy. All you have to do is take out your

electric saw and then cut along the specific lines. Here are the lines you should focus on:

- Cut along the line that connects Point A to Side A.
- Cut along the line that connects Point A to Side B.

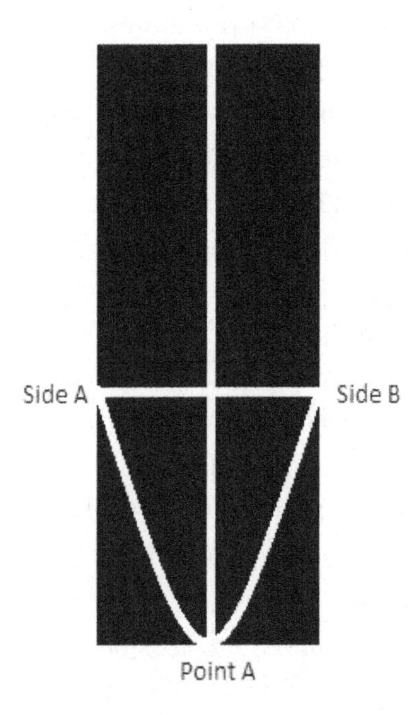

The image above represents the top view of the railroad track that you have chosen. It might or might not be a perfect rectangle, but the above design is what you should ideally aim for when making your anvil.

7. With the cuts done, it is time to subject the anvil to the grinder. You are going to smoothen the surface and give your anvil the finish that it requires.
8. If you like, you can also make use of a polish to give your anvil a nice shine.
9. With this, you have completed your anvil.

CHAPTER 9: UNDERSTANDING DIFFERENT STEELS & PROCESSES

When you are choosing a blade for forging, there are different factors that you must consider. What exactly are you going to use the blade for? Will you be using it only for cutting or also for utility purposes? Will it hold an edge for a long time? Do you want it to be flexible?

Here are some of the common metals that are used by bladesmiths:

O-1

Content

Element	Percentage of Element in Steel
Carbon	0.90%
Chrome	0.50%
Manganese	1.00%
Tungsten	0.50%

Properties

Properties	Details
Wear Resistance	Medium
Toughness	Medium
Red Hardness	Low
Distortion in Heat Treat-ing	Very Low
Forging	Start at 1,800 to 1,950°F
Quench	Oil
Tempering	350 to 500°F

0-1 is a pretty standard carbon steel for forging almost all kinds of blades, except the larger ones (such as swords). It has been known to be quite forgiving when subjected to warping or heat treatment. This makes the metal easy to work with during the heat treatment process.

W-1

Content

Element	Percentage of Element in Steel
Carbon	0.60 - 1.4%

Properties	Details
Wear Resistance	Medium
Toughness	Medium
Red Hardness	Very Low
Forging	Anywhere from 1,800 to 1,900°F
Hardening	Anywhere from 1,400 to 1,550°F
Quench	Water or Oil
Tempering	Anywhere from 350 to 650°F

This steel is a general-purpose steel where you can get a uniform depth because of its high level of hardness.

WHC

Properties

Element	Percentage of Element in Steel
Carbon	0.75%

Properties	Details
Wear Resistance	Medium
Toughness	High Medium
Red Hardness	Very Low
Forging	Anywhere from 1,850 to 1,900°F
Hardening	Anywhere from 1,400 to 1,550°F
Quench	Water or Oil
Tempering	Anywhere from 350 to 650°F

WHC has a slightly lower carbon content that W-1. However, what it lacks in the carbon department, it makes up for in the shock resistance department, with its ability to withstand shocks better than W-1.

10-SERIES STEELS

Properties of 1050

Element	Percentage of Element in Steel
Carbon	0.48 - 0.55%
Manganese	0.60 - 0.90%

Properties of 1060

Element	Percentage of Element in Steel
Carbon	0.55 - 0.65%
Manganese	0.60 - 0.90%

Properties of 1070

Element	Percentage of Element in Steel
Carbon	0.65 - 0.75%
Manganese	0.60 - 0.90%

Properties of 1080

Element	Percentage of Element in Steel
Carbon	0.75 - 0.88%
Manganese	0.60 - 0.90%

Properties of 1095

Element	Percentage of Element in Steel
Carbon	0.90 - 1.03%
Manganese	0.30 - 0.50%

Common Properties Shared by the Metals in 10-series

Properties	Details
Wear Resistance	Medium
Toughness	High to Medium, based on the carbon content

Red Hardness	Very Low
Forging	Anywhere from 1,750 to 1,850°F
Hardening	Anywhere from 1,400 to 1,550°F
Quench	Oil
Tempering	Anywhere from 300 to 500°F

When people think about which are the usable alloys for bladesmithing they are referring to the 10-series. What makes them ideal for making blades is that they are pretty stable and you can shape them easily under the hammer as compared to many other forms of steel.

5160

Properties

Element	Percentage of Element in Steel
Carbon	0.56 - 0.64%
Chromium	0.70 - 0.90%
Manganese	0.75 - 1.00%
Phosphorus	Maximum of 0.035%
Silicon	0.15 - 0.35%

Sulphur	Maximum of 0.04%

Properties	Details
Wear Resistance	High
Toughness	High
Red hardness	Low
Forging	Beginning from 1,800°F
Hardening	Anywhere from 1,450

Also popularly known as 'spring steel,' 5160 has high durability and excellent toughness. It is a medium carbon steel well suited for axes, swords, and other large blades.

L-6

Properties

Element	Percentage of Element in Steel
Carbon	0.70 - 0.90%
Chromium	0.03%

Manganese	0.35 - 0.55%
Nickel	1.4 - 2.6%
Phosphorus	0.025%
Silicon	0.25%
Vanadium	0.15%

Properties	Details
Wear Resistance	Medium
Toughness	Very High
Red Hardness	Low
Forging	Anywhere from 1,800 to 2,000°F
Quench	Oil
Hardening	Anywhere from 1,450 to 1,550°F
Tempering	Anywhere from 300 to 500°F

This steel is ideal if you are working with large saw blades and other similar items with broad blades.

S-1

Properties

Element	Percentage of Element in Steel
Carbon	0.50%
Chromium	1.50%
Tungsten	2.50%

Properties	Details
Wear Resistance	Medium
Toughness	High
Red Hardness	Medium
Forging	Anywhere from 1,850 to 2,050°F
Hardening	Anywhere from 1,650 to 1,750°F
Quench	Oil
Tempering	400 to 450°F

This steel is designed to resist long-term effects of wear and abrasion. In fact, it has been formed to absorb shocks. Its main purpose is making pneumatic and hand tools for riveting and chipping.

S-5

Properties

Element	Percentage of Element in Steel
Carbon	0.55%
Manganese	80%
Molybdenum	0.40%
Silicon	2%

Properties	Details
Wear Resistance	Medium
Toughness	Very High
Red Hardness	Medium
Forging	Anywhere from 1,650 to 1,800°F
Hardening	Anywhere from 1,600 to 1,700°F
Quench	Oil

Tempering	350 to 450°F

You use the S-5 for the same tools as the S-1. However, the S-5 is much tougher because of the molybdenum content.

Let us now look at the different processes:

NORMALIZING

The process of forging is very stressful on steel. The repeated cycles of heating and cooling, along with the physical rearrangement of the metal, creates havoc on the grain structure. Large or irregular grains can create weaknesses in your blade. This can lead to shattering or breaking during the process of making your knife or later on while the blade is being used. By normalizing your steel, you can press the reset button on your steel's grain and ensure this doesn't happen.

In fact, one of the most important parts of this is the fact that you can use it if you have made any mistakes that you would like to rectify.

In this process, you heat the metal to a particularly high temperature and then take it out to air cool. The metal gradually returns to room temperature.

ANNEALING

Some people wonder whether annealing is even necessary in the forging of blades. It is.

There are two main reasons for annealing:

- The first is to soften it and remove stress on the blade.
- The second is to make the structure of the blade even.

Every time a piece of metal is worked it accumulates stress and gets harder. The harder it gets, the more difficult it is to work again. But how exactly is annealing done?

The entire process is simple.

The metal is heated up, held at temperature for a time, and then it is slow cooled. Usually, when someone says that the metal has to be slow cooled, they are referring to the fact that it is left to cool inside the forge itself, allowing the metal to slowly return to room temperature.

Wait a minute!

Doesn't the annealing process sound suspiciously similar to the normalizing process? Can't we just call them annealing and well, annealing again?

Though it might be tempting, there is a difference.

You see, in annealing, you leave the metal in the forge. This means that it takes longer for the metal to cool down. In the normalizing process however, you are subjecting the metal to air at room temperature.

Does it make a difference?

Absolutely it does!

The main reason we anneal a metal (or in other words, allow it to cool slowly) is because of the following reasons:

- You increase the strength of the metal. You really don't want your knife breaking when all you did was use it to cut some zucchini! Annealing adds that extra durability to your knife.
- It improves the metal's ductility. A metal's ductility decides how much it can stretch before it breaks. Most knives have the capacity to bend a little to absorb shock. If they were always rigid, then they would wear easily. The best way to understand this is by using the example of our legs. When you jump from a certain elevated platform and land on the ground, you bend your knees slightly to absorb the shock and distribute it evenly throughout your body. If you do not bend, then the shock becomes focused on one region, leading to injuries. The same concept applies to knives as well. If they can bend slightly, then they have the ability to absorb more shocks. But

don't worry, it does not mean that your knife is going to behave like a limp piece of cloth.

- You can easily elongate the knife without worrying about degrading the quality. Sometimes, when you want to make the blade a little bit longer, you might end up creating cracks on the surface. To avoid such scenarios, we subject the metal to the annealing process.

QUENCHING

This is a pretty popular process in metallurgy and you might have seen it being performed on TV or in the movies. Basically, the metal worker takes the steel and then dips it into a cauldron, often creating a dramatic effect. You can see steam rising and the water sizzling as the metal cools. However, water is not the only medium that the metal worker uses for the process of quenching. Before we dive into that, let us look at the quenching process in more detail.

Quenching is a process that occurs after another process where the metal is heated to high temperatures. Examples of such processes preceding quenching are annealing (which we looked at earlier) or normalizing (which we shall look at in the next section). In both annealing and normalizing, the cooling process can take some time. This could affect the strength of the steel, causing it to lower more than necessary. Through quenching, you are lowering the temperature considerably, which could benefit the work that you are doing. Metal workers usually apply this method so that they can prevent the cooling process from altering the molecular structure of the metal.

Quenching is typically done by submerging the metal immediately to a certain liquid, typically water, or forced air. In a forced air-cooling system, the air is pushed out through specially arranged ducts that help in, you guessed it, cooling the temperature of an object quickly. The water or air used for the process of quenching is termed the 'medium.'

Now you might think to yourself: Is there any other liquid that can be used for quenching? That is actually a valid point. We are so used to seeing the movies show metallic objects such as swords and weapons being dipped into water that we are not aware of any other liquids that can be used. However, here a couple of other liquids that are used for quenching:

Oil

There are numerous oil options that you can use for the process of quenching. You have fish oils, vegetable oils, and certain mineral oils that can help you attain the desired effect. With each medium, you have a different rate of cooling. When you use oil, you are using a liquid that has a higher cooling rate than air but cools the metal down more slowly than water.

When choosing oils, here are a few options that you can consider for your bladesmithing requirements:

Food Grade Oils

Many bladesmiths and knifemakers utilize food grade oils for the sole reason that they are cheaper and readily available. You can head over to your local supermarket or store and find some on their shelves. In fact, you are also spoilt for choice, with the number of different brands available to you. One famous food oil used in quenching is canola oil.

During the quenching and tempering processes, food oils spread a much more bearable odor than other types of oil. While you might think that this is a minor point to make note of, it might affect you if you have a workshop attached to your home or within your home.

To quench in food grade oils, preheat the oil to anywhere between 150°F and 200°F.

Motor Oil

Another oil that is popularly used in the knifemaking industry or among knifemaking hobbyists is motor oil. People use both new and used motor oil, depending on their requirements. The advantage of motor oil is that it is really cheap to obtain. In fact, used motor oil is free; you may have it in your garage, you might find some in your friend's garage, or even in the local store.

However, do note that used motor oil tends to leave a stench. Additionally, you might find a dark film coating the blade you are working on, which is quite difficult to remove. Another reason why you might want to stay away from used motor oil is the fact that it contains quite a few toxins that could be potentially harmful if inhaled. For beginners, it is highly recommended that they do not use motor oils. Experienced knife makers who have worked with the blade for many years might be able to spot good motor oil. Nevertheless, they prefer to avoid it as well if they have a choice.

To quench in motor oils, preheat the oil to anywhere between 200°F and 250°F.

Mineral Oil

Mineral oil is an alternative to motor oil. Its benefits are that it does not give off any odor, does not contain any harmful contaminants or toxins, and some of them are fairly odorless. If you can get your hands on high-grade mineral oil, then you might be able to avoid the flames that flare up during the quenching process.

In many cases, mineral oil is recommended for beginners.

To quench in mineral oil, preheat the oil to anywhere between 250°F and 300°F.

Baby Oil

Yes, you heard that right: baby oil. Many knifemakers head over to their local store and get a lot of baby oil for the process of quenching. Baby oil contains minimal contaminants, does not create flames easily, and tends to give rise to as little odor as possible. Should you prefer, you could even get yourself some scented baby oil. You can quench your blade and leave behind a nice fragrance as an extra-added bonus!

The downside to that is that you might require a fair amount of baby oil, and you might find yourself shelling out a fair bit of cash to acquire the right volume of oil for your project.

You can use the same preheating temperature for baby oil as you did for food oil, Aim for temperatures between 150°F and 200°F.

Quenching Oils

Finally, you can purchase special quenching oils for your project, taking advantage of the fact that you can find the right oil for a specific purpose. Working on a blade? You have a quenching oil for that. Working on an anvil? Sure, there is a quenching oil for that too. Need a specific quenching oil for a specific steel? Still, no problem.

Of course, with the number of options available to you, quenching oils might be a bit more expensive than other types of oils.

For quenching oils, the quenching temperature is unique to the type of oil you choose to purchase. The quenching temperature will be mentioned on the packaging, allowing you to not only have instructions for quenching but also to help you decide just what quenching oil you would like to use for your project.

One of the most popular fast quenching oils that you can find on the market is Parks 50.

FORGING STEEL COLORS

Fahrenheit	The Color of the Steel	Process
2,000°	Bright Yellow	Forging
1,900°	Dark Yellow	Forging
1,800°	Orange Yellow	Forging
1,700°	Orange	Forging
1,600°	Orange Red	Forging
1,500°	Bright Red	Forging
1,400°	Red	Forging
1,300°	Medium Red	-

1,200°	Dull Red	-
1,100°	Slight Red	-
1,000°	Mostly Grey	-
800°	Dark Grey	Tempering
575°	Blue	Tempering
540°	Dark Purple	Tempering
520°	Purple	Tempering
480°	Brown	Tempering
445°	Light Straw	Tempering

One of the things that you will notice when working with metals is that you usually aim for a color that is obvious enough as to exactly which color you are looking at. For example, if you heat the metal to orange, then the temperature of the metal is anywhere between 1,600 and 1,800°F. When you reach the temperature range, then you are good to proceed with the next step in the process.

CHAPTER 10: FINISHING THE KNIFE

GRINDING

Grinding is the process of using files or sanding machines to shape the steel into forming a bevel.

Many people become concerned about the fact that the heat produced during grinding might affect the knife. This is not true. You can breathe a sigh of relief. The main reason for this is that you are typically grinding the knife before subjecting it to any form of heat treatment processes.

I understand that other bladesmiths prefer grinding after heat treatment. But in my opinion, doing it before heat treatment has less risks, especially when dealing with scrap steel.

There are many types of grinds that you can perform on your knife. But we are going to do a full flat grind on your knife. This guide has a few pointers on doing a flat grind on your knife. Let's get started.

1. The first thing that you are going to do is take out your knife and bring it close to the grinder. You should preferably use a fresh belt when you are grinding your knife so that it does not heat up the blade too much. The sequence of grits for the

belts you should be using should be 36,60,120, and 220.

2. Do not press the blade too much towards the grind. Gently touch it and start moving the knife. This will allow the grind to make the shape of the edge on the knife.
3. Start the grind close to the edge and slowly make your way up to the spine. You will notice that the edge has a gradual and linear slope like shape to it now. Once you have worked on one side of the knife, flip it over.
4. Bring it close to the grind and then add the slope like shape on the other side as well.

MAKING THE FULL TANG KNIFE HANDLE

A suitable piece of handle material must be chosen for this process. You can choose a natural material, usually an exotic wood, or a man-made material such as Micarta. We are using wood as the handle material in this set of instructions. The wood will need to be cut down the middle to make two halves. This can be done with the bandsaw or a handsaw. The handle design should be marked on the outer part of the scales, with a pencil to avoid confusion later. It is important that the grain of the wood lines up as it did before it was cut. The inside cuts are ground smooth and flat with one of the belt sanders.

The forward edge of one of the handle pieces is ground square and then held against the guard and viewed in the light. If there are any gaps in the guard/handle joint, they will be visible when examined in the light.

1. Cut off the handle material to close to the shape of your markings carefully. You are doing this because you want the material to match the tang design of the knife.
2. Clamp together the scales and the tang before drilling, so that the holes are in line.
3. Drill the holes into the tang as marked by the pencil design. Then cut out rivets or use Corby fasteners as pins for the handle. Hammer them in place inside the holes and make sure they fit tight.
4. Rub the scales using acetone to remove any moisture from them.
5. Clean the tang and guard with acetone to remove any oils or dirt, and then mix some epoxy. The first side of handle material is carefully glued and clamped in place, making sure that the handle joint is tight. Once the glue has set, the pinholes are hammered through the handle from the side of the tang. If the handle material does not follow the profile of the tang, just take a bit of time to cut it closer to the final dimensions.
6. If there is too much epoxy, then you have to make sure that it gets cleaned off to allow it to attach sufficiently the other side of the handle. The forward end of the remaining handle piece is mated to the guard using the light to determine needed adjustments. Then the exposed tang is cleaned

again with acetone, new epoxy is mixed, and the remaining handle is glued to the knife.

7. The handle is now shaped using the belt sander. Pinholes are left open until the handle is sanded to #120 grit. Then the pins are applied a little bit of epoxy and added, peened in place if necessary and grounded flush with the handle.

SHARPENING

Sharpening the edge of your blade is the last step in making your knife. Like most things in knifemaking, every maker has their preferred method for sharpening. You should ideally look for a sharpening process where you are able to use the knife effectively for your lifestyle and needs.

So how to sharpen the blade?

There are many ways but here are some that might be useful for you.

Apart from the sharpeners mentioned in the list below, you can also use the Lansky Knife Sharpening System. The Lansky is a complete kit for sharpening that is ideal for beginners.

V Sharpener

These sharpening tools have two edges, shaped in a V. A blade can be moved through the V with light pressure to grind the edge. By holding the sharpeners at a specific angle, it creates consistency in every pass that can be hard to obtain using a stone. Unfortunately, it most likely isn't the exact angle you want. It also won't compensate for the loss of steel that occurs with use over time and will change the shape of your edge.

Waterstones

Waterstones require water to be used as a lubricant. One of the things you might notice when using them is that their surface keeps wearing away. This exposes new layers and makes them great for sharpening. You can also use this stone to polish the bevel. The constant wear creates an indentation in the top of the stone, which needs to be flattened frequently to maintain a good surface for grinding. The waterstone is an option that is recommended for beginners. It allows beginners to easily grasp the basics of sharpening.

Belt Grinder

Belt grinders are one of the most popular forms of sharpeners and as you might have noticed, we have used the grinding process liberally throughout this book. Basically, the same grinder that you used to make your grinds can be used to sharpen your edge. The knife is held gently at the appropriate angle against the slack part of a fine belt. Doing this takes a good eye and a steady hand and can be difficult for a beginner due to the speed at which the metal is removed. Grinding in this manner tends to create a slightly convex grind.

TAKING CARE OF YOUR KNIFE

After you have invested a lot of time into your blade, it would be a shame if you were to ruin it too quickly.

So here are a few notes that you should remember about knife care.

Those who make knives are at an advantage when it comes to knowing how to take care of a blade. Now that you know what it takes to make your blade, it's easier to focus on avoiding the situations that what will ruin one. If you find yourself in a survival scenario, it's incredibly vital to protect the knife that you are going to use to keep you alive. Even if you are not, it can be harmful to ruin something you have taken so much time to create.

One of the things that ruin the knife quicker is when people place it in the dirt whenever they are outside. The idea behind that is that they can easily reach for their knives when they have to. But this does not do anything except dull the blade of the knife even further. Of course, it does look cool in the movies when the actor does it. But when have we ever used movies as a guide to real life situations?

Avoid using any cutting surface that is as hard as your blade's edge. Some people try and use a rock as a cutting surface. Don't do that. Your blade will be grateful to you if you use a log or piece of wood instead. Cutting boards not only serve as a surface to cut on, but also are also important to protect the integrity of the knife.

Make sure to dry your blade when it gets wet and to rinse it if it comes into contact with saltwater. Don’t store your blade for long periods of time in the sheath, as the leather can collect moisture and cause corrosion.

All carbon steel will need care to prevent corrosion in the form of surface rust, so anything you can do to be proactive is a big help. If you do see rust forming, use fine grit paper to remove it as soon as possible and apply a light coat of oil.

You can also get a patina on the knife instead of rust. Now the line of difference between a patina and rust is very thin as they might almost look similar. However, patina is more controlled whereas rust can spread easily. In order to create a patina, simply dip your blade into boiling water. This oxidizes the rust and you have a fine layer of patina on your knife.

CONCLUSION

Forging is not difficult.

However, for many of the processes, the results won't be instant. You have to slowly work your way to getting the desired results.

And that is where more people fail. That is what makes forging difficult; it is not the idea that people can't get something done, but the fact that they can't get it done instantly. It is a laboring process and it does tax your body as well.

However, remember that as long as you make sure that you are taking precautions at all times to protect yourself, work with the instructions you have, then you won't have to worry about anything.

With that, I hope you enjoy forging and creating your own very fine blade.

REFERENCES

Hrisoulas, J. (2010). *Master bladesmith*. Boulder, Colo.: Paladin.

Hrisoulas, J. (2017). *Pattern-Welded Blade*. NE-PHILIM Press.

Hrisoulas, J., Morris, J. and Sherbring, M. (2017). *The complete bladesmith*. Redd Ink Press.

Sims, L. (2009). *The backyard blacksmith*. Crestline Books.

Made in the USA
Coppell, TX
19 October 2021